U0171236

我怎么能够

把你来比作夏天

365 Days of Love

爱情手账

[英] 多尔顿·埃克斯利

——编选

周成刚——

——主编

新星出版社 NEW STAR PRESS

Every day is a new day.

每一天，都是新的开始。

Loving someone and having them love you back is the most precious thing in the world.

世上最珍贵的莫过于爱与被爱。

——尼古拉斯·斯帕克思（NICHOLAS SPARKS）

Time and love, those are the greatest gifts of all.
世上最美好的礼物是时光与爱。

——托尼·霍克（TONY HAWKS）

That is true love which always and forever remains the same, whether one grants it everything or denies it everything.

无论你全然接受还是全盘否定，真爱永恒不变。

——约翰·沃尔夫冈·冯·歌德（JOHANN WOLFGANG VON GOETHE）

Eventually you will come to understand that love heals everything, and love is all there is.

世人终将领悟：爱，治愈一切；爱，即是全部。

——加里·祖卡夫（GARY ZUKAV）

Love is the greatest thing in the world.
爱是世间最伟大的力量。

——奥斯卡·王尔德（OSCAR WILDE）

Only here, only in each other's arms, we rediscover joy.
Only here we are ourselves, and so each other's.

只有在彼此的怀抱里，我们才能重拾欢乐。
此时此刻，回归自我，也找回彼此。

——帕姆·布朗（PAM BROWN）

7

Date:　　/　　/

Mon.	Tue.	Wed.	Thu.	Fri.	Sat.	Sun.

I want you for always – days, years, eternities.
我愿与你相依相伴——每一天、每一年，直到永远。
——罗伯特·舒曼（ROBERT SCHUMANN）

No one knows our secret. We seem such an ordinary couple. How could they know the depth and wonder of our love?

我们的秘密无人能懂，我们是如此平凡的爱侣，可谁人知道我们爱之深、情之切？

——布赖恩·威廉姆斯（BRIAN E. WILLIAMS）

The key to love is faith — and being able to trust those whom we love and who love us.

忠贞使我们恒久相爱，彼此信任。

——斯图尔特·瑞林格（STUART REININGER）

All I have is your love. I have everything.
你的爱就是我的全部，我因你的爱而拥有了整个世界。
——斯图尔特与琳达·麦克法兰（STUART & LINDA MACFARLANE）

Date:　　／　　／

Mon.	Tue.	Wed.	Thu.	Fri.	Sat.	Sun.

Love is not consolation, it is light.

爱不是慰藉，爱是光亮。

——西蒙娜·薇伊（SIMONE WEIL）

Fair is the white star of twilight, and the sky cleaner at the day's end;
But she is fairer, and she is dearer, She, my heart's friend !
薄暮时分，繁星闪耀，天空澄净。此时的她，愈加美丽，愈加可贵。
她是我的心之所属。

——肖肖尼情歌（SHOSHONE LOVE SONG）

All, everything that I understand, I understand only because I love.
拥有了爱，我便领悟了万事万物的真谛。

——列夫·托尔斯泰（LEO TOLSTOY）

Mon.	Tue.	Wed.	Thu.	Fri.	Sat.	Sun.

If the soul is to know itself, it must look into a soul.

若要透视自己的灵魂深处，须与伴侣之灵魂碰撞出爱的火花。

——乔治·塞菲里斯（GEORGE SEFERIS）

Date:　/　/

Mon.	Tue.	Wed.	Thu.	Fri.	Sat.	Sun.

When it comes down to it, we all just want to be loved.

归根结底，我们都希望被人爱着。

——杰米·耶林（JAMIE YELLIN）

Beauty is a light that shines through the most ordinary woman – if she is happy and in love.
只要拥有快乐与爱，即便是最寻常的女子，她也会美得夺目。

——帕姆·布朗（PAM BROWN）

The disease of love has no physician.

爱情的烦恼无医可解。

——斯瓦希里情诗（SWAHILI LOVE POEM）

Gravity may hold us to the planet but love holds us together.
地心引力也许会让我们留在地球上，而爱情让我们拥有彼此。
——斯图尔特与琳达·麦克法兰（STUART & LINDA MACFARLANE）

Riches take wings, comforts vanish, hope withers away, but love stays with us.

财富会消散，慰藉会消退，希望会幻灭，但是爱永远与我们同在。

——刘易斯·华莱士（LEWIS "LEW" WALLACE）

We never leave each other. When does your mouth say goodbye to your heart?

我们彼此永不分离，正如我们永远不会和自己说再见。

——玛丽·托芒顿（MARY TALL MOUNTAIN）

The tender words we said to one another are stored in the secret heart of heaven:
One day like rain they will fall and spread, and our mystery will grow green over
the world.

我们彼此说过的最温柔的话语是我们存在这欢乐世界里的秘密，总有一天，这
秘密会像雨滴一样落下漫延，而我们的传奇会在这世界生根发芽，生生不已。

——札拉丁·鲁米（JALAL AL-DIN RUMI）

Here's the scarf you left behind holding the scent of you.
I press it to my face, remembering.
你把围巾留给了我，上面有你的芳香，我把围巾贴在脸上，想你。

——夏洛特·格雷（CHARLOTTE GRAY）

23

Without love, we are birds with broken wings.
若无爱，我们便是折翼的鸟儿。

——莫里·施瓦茨（MORRIE SCHWARTZ）

Date: /   /

Mon. Tue. Wed. Thu. Fri. Sat. Sun.

24

Those who have never known the deep intimacy and hence the companionship of happy mutual love have missed the best thing that life has to give.
若从未拥有亲密关系，并因此缺乏爱人的幸福陪伴，便已然错过了生命中的至美。

———伯特兰·罗素（BERTRAND RUSSELL）

You are never too young to fall in love and never too old to wish you had.

坠入爱河，二八芳龄亦有之；追逐爱情，年届花甲终不晚。

——克里·诺布尔（KERI NOBLE）

Everything that touches us, me and you, takes us together like a violin's bow, which draws one voice out of two separate strings.

我们共同经历的一切将我和你紧紧地连在一起，我们宛如提琴的琴弦，琴瑟和鸣。

——莱纳·玛利亚·里尔克（RAINER MARIA RILKE）

What comes from the heart, touches the heart.

发乎心者，触动心弦。

——唐·西柏（DON SIBET）

You can do anything and be anything, so long as it's with me.
只要有我相伴，你便可追逐梦想、如愿以偿。

——迪伦·托马斯（DYLAN THOMAS）

Being with you is like walking on a very clear morning – definitely the sensation of belonging there.

和你相依相伴，就如在明朗的清晨漫步，让我心有所属。

——埃尔文·布鲁克斯·怀特（E. B. WHITE）

When problems arise, we tend to forget about the strong foundation we have in our relationship. Remember where and when you and your partner felt the strongest, the closest, and the most intimate. When you are at the lowest point of your relationship, you can have that as a frame of reference.

当发生争执时，我们常常会忘记我们曾经的深情，请记住你和爱人的亲密时光吧，当我们沮丧的时候，就多想想那些幸福快乐的瞬间吧。

——索邦福索（SOBONFU SOMÉ）

The entire sum of existence is the magic of being needed by just one person.
我们这一生哪怕只被一人所需，这神奇的魅力便是生命存在的全部意义。
——维帕特南（VI PUTNAM）

When you remove love and try to replace it with monetary things, you've got nothing.

当你试图用金钱取代爱情时，你终将一无所获。

——约翰·彼得斯（JOHN PETERS）

Never above you. Never below you. Always beside you.
爱，非睥睨，非仰视。爱，比翼齐飞。
——沃尔特·温切尔（WALTER WINCHELL）

Love, whereby two people walk in different directions yet always remain side by side.

即使我们已经相向而行，而爱终将会让我们并肩前行。

——休·普拉瑟（HUGH PRATHER）

Happiness is being snug abed with the person you love – and the rain lashing the window and drumming on the roof. Safe.

幸福就是执子之手，相依相守。观斜风细雨，听雨打芭蕉，内心安宁。

——帕姆·布朗（PAM BROWN）

It is the true season of love when we know that we alone can love, that no one could ever have loved before us and no one will ever love in the same way after us.

我唯有和你相恋，才是真正的爱恋时刻，没有人曾拥有过我们这样的深情，将来也没有人可以超越你我的爱情。

——约翰·沃尔夫冈·冯·歌德（JOHANN WOLFGANG VON GOETHE）

He who defends with love will be secure.
爱以身为天下，若可托天下。

——老子（LAO TZU）

...if you love someone, you need to be with them, close to them. You need to be able to confide, to laugh together. It's just about as important as breathing.

若爱上一个人，需时常相伴左右，关怀无微不至，相互信任，同甘共苦，如同呼吸，必不可少。

——罗莎蒙德·皮尔彻（ROSAMUNDE PILCHER）

Love has nothing to do with what you are expecting to get – only what you are expecting to give – which is everything.

爱是给予，而非索取，这就是爱的一切。

——凯瑟琳·赫本（KATHARINE HEPBURN）

Don't ask me to leave you! Let me go with you. Wherever you go, I will go; wherever you live, I will live.

请别让我离开你，我要与你相依相随，走遍千山万水。

——《圣经·旧约》（OLD TESTAMENT）

41

I am nearly mad about you as much as one can be mad; I cannot bring together two ideas that you do not interpose yourself between them. I can no longer think of anything but you.

我为你痴迷，为爱痴狂。我不能把你和他人同等对待，我的心里只有你。

——斯图尔特与琳达·麦克法兰（STUART & LINDA MACFARLANE）

Date: / /

| Mon. | Tue. | Wed. | Thu. | Fri. | Sat. | Sun. |

When two people fall in love, each comes in out of the loneliness of exile, home to the one house of belonging.

当我们相爱，我们从此不再孤独，我们的心不再流浪，我们拥有了彼此。

——约翰·奥多诺休（JOHN O'DONOHUE）

Love is an act of endless forgiveness, a tender look which becomes a habit.
爱是无尽的宽容，爱是永远温柔的目光。

——彼得·乌斯蒂诺夫（PETER USTINOV）

Time is too slow for those who wait, too swift for those who fear, too long for those who grieve, too short for those who rejoice, but for those who love, time is eternity.
于等待者而言，时间太慢；于恐惧者而言，时间太快；于悲伤者而言，时间太长；于欢乐者而言，时间太短；然而，于沐浴爱河者而言，时间永恒。

——亨利·凡·戴克（HENRY VAN DYKE）

45

My heart has made its mind up
And I'm afraid it's you.
我已心有所属……恐怕那个人就是你。

——温迪·科普（WENDY COPE）

Security is when I'm very much in love with somebody extraordinary who loves me back.

当我爱上一个非凡的人，而他也爱我，这就是爱情带给我们的安全感。

——雪莉·温特斯（SHELLEY WINTERS）

Love never dies. It may look different, take on different shapes, ebb, flow, flicker, and blaze, but it is the one thing in this world that never, ever dies.

爱永恒。爱有千姿百态，或起伏跌宕，或百转千回，或时隐时现，或熠熠生辉。世间唯爱永存。

——凯拉·默温（KYLA MERWIN）

My lifetime listens to yours.
我每时每刻都要倾听你的心声。

——缪里尔·鲁凯泽（MURIEL RUKEYSER）

Love is supposed to start with bells ringing and go downhill from there. But it was the opposite for me. There's an intense connection between us, and as we stayed together, the bells rang louder.

人们普遍认为，自婚礼教堂钟声响起之日起，爱情便逐渐暗淡无光。然于我而言，恰恰相反。你我爱得刻骨铭心、相依相伴，爱情钟声更加嘹亮。

——莉萨·尼米（LISA NIEMI）

For all the little, unexpected surprises. For all the enduring kindness –
thank you.
感谢你的小小惊喜，感谢你的深情厚谊。

———帕姆·布朗（PAM BROWN）

The smallest whisper of love can restore confidence and sureness.
哪怕是最简短的轻声呢喃，只要有爱就能重拾信心，抚慰心灵。
——约翰·奥多诺休（JOHN O'DONOHUE）

Date: /   /

Mon. Tue. Wed. Thu. Fri. Sat. Sun.

To fear love is to fear life; and those that fear life are already three parts dead.
惧怕爱情就是惧怕生活，惧怕生活的人就如同行尸走肉。
——伯特兰·罗素（BERTRAND RUSSELL）

The first duty of love is to listen.
爱的首要职责在于倾听。

——保罗·田立克（PAUL TILLICH）

Date:　　/　　/

Mon.	Tue.	Wed.	Thu.	Fri.	Sat.	Sun.

54

I want to be your friend forever and ever. When the hills are all flat and the rivers are all dry, when the trees blossom in winter and the snow falls in summer, when heaven and earth mix – not till then will I part from you.

我欲与君相知，长命无绝衰。山无陵，江水为竭。冬雷震震，夏雨雪。天地合，乃敢与君绝。

——《乐府诗集》（THE YÜEH-FU）

The quiet thoughts of two people a long time in love touch lightly like birds nesting in each other's warmth. You will know them by their laughter....

长时间相知相爱的伴侣，注重心灵无言的交流，互相取暖，共筑爱巢。光凭他们的笑声你就能了解这一切。

——休·普拉瑟（HUGH PRATHER）

Love may have its ups and downs – but it's better to be in than out.
经受爱情的折磨也比远离爱情好。
——斯图尔特与琳达·麦克法兰（STUART & LINDA MACFARLANE）

A star is shining in my heart, My dreams have wings that touch the sky, I'd marry you a thousand times – I'll love you till the day I die.

心中有星星闪耀，梦想就有了飞翔的翅膀，我愿嫁你千百次，我会爱你到地老天荒。

——玛丽昂·肖伯林（MARION SCHOEBERLEIN）

…when love speaks, the voice of all the gods make heaven drowsy with the harmony.

万籁俱静，爱的呢喃，宛如天籁之音。

——威廉·莎士比亚（WILLIAM SHAKESPEARE）

All of our experiences in life make us not less valuable, but more valuable, not less able to love, but more able to love.

生活赋予我们的一切会使我们变得更有价值，更加懂得如何去爱。

——乔安娜·坎贝尔·斯兰（JOANNA CAMPBELL SLAN）

The wind whispering secrets to the trees,
The snowflakes floating on crisp clear air.
Their beauty would be lost,
Their enchantment stolen,
Should I ever have to live without your love.

微风摇曳，树影婆娑，
雪花飘落，片片含情。
如若没有你的爱，良辰美景瞬间魅力尽失。
如若没有你的爱，我的人生，亦无亮色。

——斯图尔特与琳达·麦克法兰（STUART & LINDA MACFARLANE）

Love is a net where hearts are caught like fish.
爱如一张网，捕获爱人的心。

——苏菲箴言（SUFI SAYING）

| Mon. | Tue. | Wed. | Thu. | Fri. | Sat. | Sun. |

Love doesn't just sit there like a stone; it has to be made, like bread, remade all the time, made new.

爱不是像僵硬的石头那样，一动不动地待在那里。对待爱，要像做面包那样，反复揉搓，不断翻新。

——厄休拉·K. 勒吉恩（URSULA K. LE GUIN）

Saying "I love you" is a conversation, not a message.

"我爱你"不是一句简单的信息传递，而是你来我往、永不停歇的情歌对唱。

——道格拉斯·斯通（DOUGLAS STONE）

All love is sweet, given or returned.
不论是给予还是回报，爱总是甜蜜的。

——珀西·比希·雪莱（PERCY BYSSHE SHELLEY）

65

Date: / /

Mon.	Tue.	Wed.	Thu.	Fri.	Sat.	Sun.

In the rich tapestry of life it is the bright threads of love that fashion the scenes of happiness.

在多彩斑斓的生活中，爱情，以其靓丽风姿点着幸福时刻。

——斯图尔特与琳达·麦克法兰（STUART & LINDA MACFARLANE）

You live that you may learn to love.
You love that you may learn to live.
No other lesson is required of us.
人生有两门功课需要学习——活着是为了学会去爱，爱是为了理解活着的意义。

——米哈伊尔·奈米（MIKHAIL NAIMY）

What is yours is mine, and all mine is yours.
你中有我，我中有你。

——普劳图斯（PLAUTUS）

In a full heart there is room for everything, and in an empty heart there is room for nothing.

心若归整，万事可承；心若皆空，无事能容。

——安东尼奥·波契亚（ANTONIO PORCHIA）

Your task is not to seek for love, but merely to seek and find all the barriers within yourself that you have built against it.

你需要做的不是去追寻爱情，而是找到所有在你心里生根的爱情屏障。

——札拉丁·鲁米（JALAL AL-DIN RUMI）

And when we kiss, a current surges, heart to heart, carrying all my love to him and all his love to me.

吻上你的唇，爱的激流涌上心头。你我全部的爱，在此交融。

——琳达·麦克法兰（LINDA MACFARLANE）

Age does not protect you from love. But love, to some extent, protects you from age.

岁月增长，非爱之殇。爱之滋养，年月可藏。

——让娜·莫罗（JEANNE MOREAU）

To love and be loved is to feel the sun from both sides.
爱与被爱，如沐暖阳。

——大卫·威斯科特（DAVID VISCOTT）

Anything, everything, little or big becomes an adventure when the right person shares it.

与对的人一起经历的任何事，无论大小，皆成奇遇。

——凯瑟琳·诺里斯（KATHLEEN NORRIS）

Love is never elusive. In all its permutations, love surrounds us in the world, whether we are accepting of it or not.

无论接纳还是抗拒，爱是无法逃避的，爱以它的方式陪伴我们。

——西德尼·波埃特（SIDNEY POITIER）

Loving affection for living beings is the water with which to irrigate the field of the mind and make it fertile.

爱恋如涓涓细流，滋润心田。

——索南仁钦格西（GESHE SONAM RINCHEN）

Between a man and his wife nothing ought to rule but love.
男人与妻子之间，唯爱统领一切。

——威廉·佩恩（WILLIAM PENN）

Date: /   /

Mon.	Tue.	Wed.	Thu.	Fri.	Sat.	Sun.

Affection is a coal that must be cooled, Else, suffered, it will set the heart on fire. The sea hath bounds, but deep desire hath none.

爱之烈焰，时冷却之，若其不然，心火炽之。海亦有涯，而欲壑难填。

——威廉·莎士比亚（WILLIAM SHAKESPEARE）

Come what may as long as you live, it is day.
And if I in the world must roam,
Wherever you are that is home.
When your loving voice I hear,
The future's shadows disappear.

你在身畔，便是天明。
我若游于天际，
凡你所在之处，皆尽为家。
听闻你爱的细语，
未来世界再无阴云。

——台奥多尔·施笃姆（THEODOR STORM）

Familiar acts are beautiful through love.
爱，使一切平凡的事物变得美好。

——珀西·比希·雪莱（PERCY BYSSHE SHELLEY）

To love means to say: it is good that you are you, it is very good.

爱意味着永远做自己，活成自己想要的模样。

——拉第斯劳·保罗（LADISLAAUS BOROS）

Date: / /

Mon.	Tue.	Wed.	Thu.	Fri.	Sat.	Sun.

Even the most ordinary day is made special by your love.

平凡的日子，因你的爱，也变得不平凡。

——帕姆·布朗（PAM BROWN）

Sally has a smile I would accept as my last view of earth.
萨莉的微笑，成为我此生永恒的记忆。

——华莱士·斯特格纳（WALLACE STEGNER）

Love doesn't attempt to bind, ensnare, capture. It is light, free of the burden of attachments. Love asks nothing, is fulfilled in itself.

爱不是束缚，不是诱惑。爱是轻盈的，爱是自由的。爱不需要索取，爱自会给予我们幸福美满。

——维马拉·塔萨尔（VIMALA THAKAR）

Love cures people, both the ones who give it, and the ones who receive it.
爱与被爱皆为治愈。

——卡尔·门宁格（KARL MENNINGER）

I love you soulfully and bodyfully, properly and improperly, every way that a woman can be loved.

我的灵魂和身体都深爱着你，我倾尽所有，只为爱你。

——萧伯纳（GEORGE BERNARD SHAW）

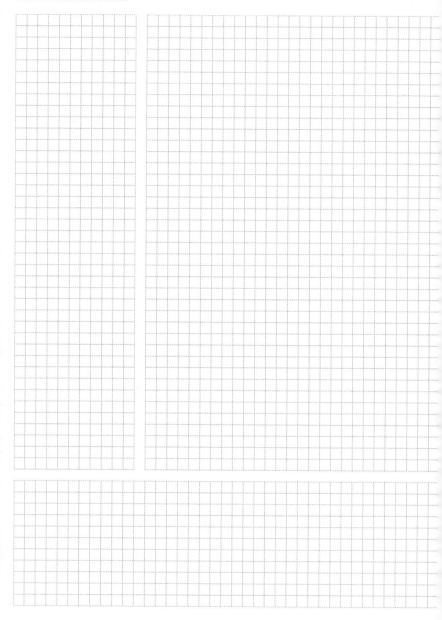

In love all of life's contradictions dissolve and disappear. Only in love are unity and duality not in conflict.

因为爱，生命中不再有对立；因为爱，一切都变得友爱、和谐，世界不再有冲突。

——拉宾德拉纳特·泰戈尔（RABINDRANATH TAGORE）

My bounty is as boundless as the sea, My love as deep; the more I give to thee, The more I have, for both are infinite.

恩泽若海宽，爱情似洋深；因爱无限，给予越多，拥有越多。

——威廉·莎士比亚（WILLIAM SHAKESPEARE）

Wherever you are, I am there also.
无论你走到天涯海角，我的精神都与你同在。
——路德维希·凡·贝多芬（LUDWIG VAN BEETHOVEN）

Most certain and more sure – the heartbeat that marks time to all I do – my second self and yet uniquely you. Nothing can overwhelm me utterly, so long as you are here. Distance can never separate us. Time can only bring us closer.

我愈加确信：令我心跳不已的是——我的另一半、独一无二的你。只要你在身旁，我便无所畏惧。距离虽远，吾心永恒；时间流逝，爱情弥深。

——帕姆·布朗（PAM BROWN）

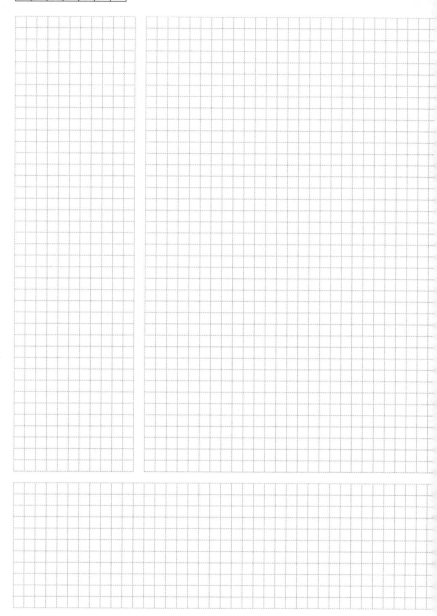

What happiness to be beloved; and O, what bliss, ye gods, to love!
被爱是何等的幸福啊，神啊，坠入爱河是何等幸事啊！
——约翰·沃尔夫冈·冯·歌德（JOHANN WOLFGANG VON GOETHE）

If this is love... can I have seconds?

如果这就是爱情……我可否享有片刻？

——斯图尔特与琳达·麦克法兰（STUART & LINDA MACFARLANE）

So, either by thy picture or my love,
Thyself away thou art present still with me;
For thou not farther than my thoughts canst move,
And I am still with them, and they with thee…

因此，看着你的照片或是凭借着我的爱，
你仍然与我在一起；
你并不会比我的思想走得更远，
而我仍然与这些回忆同在，它们也与你同在……

——威廉·莎士比亚（WILLIAM SHAKESPEARE）

93

Date: /   /

Mon.	Tue.	Wed.	Thu.	Fri.	Sat.	Sun.

Those who love deeply never grow old; they may die of old age, but they die young.

深爱着的人永远不会老。他们可能因年事已高而离去，但辞世时却依然年轻。

——亚瑟·温·皮内罗爵士（SIR ARTHUR WING PINERO）

Of all forms of caution, caution in love is perhaps the most fatal to true happiness.
我们或许过着谨慎的生活，但对爱情谨慎也许是获得真正幸福的最致命障碍。

——伯特兰·罗素（BERTRAND RUSSELL）

When you are touched by love, it reaches down into your deepest fibre.
When you are loved, your heart rushes forth in the joy of the dance of life.
Love awakens the youthfulness of the heart.

当你被爱情感动时，爱情会触及你内心最深处。当你被爱时，你的心
在生命之舞的欢乐中奔涌而出。爱情唤醒心灵的青春。

——约翰·奥多诺休（JOHN O'DONOHUE）

Love is a wonderful thing. It is more precious than emeralds, and dearer than fine opals. Pearls and pomegranates cannot buy it, nor is it set forth in the market-place. It may not be purchased of the merchants, nor can it be weighed out in the balance for gold.

爱情很美妙。它比绿宝石更珍贵，比上乘的猫眼石价更高。珍珠玛瑙换不回，市场上也无处可寻，无处购买，也无法用称黄金的天平来计量。

——奥斯卡·王尔德（OSCAR WILDE）

Alone in a crowd, I can feel isolated in secret joy just thinking of him.
我伫立在人群中，孑身一人，却因想到他，而内心暗自欢喜。

——N. 佩恩（N. PAYNE）

There is a comfort in the strength of love;
Twill make a thing endurable, which else
Would overset the brain, or break the heart.

爱情的力量可以抚慰人心。
有了爱情，我们可以变得持久忍耐，
可如若没有爱情，
我们会变得焦躁不安、心碎不已。

——威廉·华兹华斯（WILLIAM WORDSWORTH

Love is totally unpredictable. You never know where it is going to turn up and in what form. And you can never have enough of it. Love is a state of grace.

爱情是不可预期的。你永远不知道它会以何种形式、在何处出现。我们永远渴望爱情，爱情是恩典。

——凯西·莱特（KATHY LETTE）

The ultimate test of a relationship is to disagree but to hold hands.
一段亲密关系的终极考验是：意见相左，但携手前行。
——亚历山德拉·彭尼（ALEXANDRA PENNEY）

Date: / /

Mon.	Tue.	Wed.	Thu.	Fri.	Sat.	Sun.

What's mine is yours, and what is yours is mine.
我的就是你的，你的就是我的。

——威廉·莎士比亚（WILLIAM SHAKESPEARE）

The friendship between us, the mutual confidence, the delights of the heart, the enchantment of the soul, these things do not perish and can never be destroyed. I shall love you until I die.

我们之间的友谊、彼此的信任、内心的喜悦、灵魂的魅力，所有这些都不会消逝，也永远不会被摧毁。我将爱你直至我生命的尽头。

——伏尔泰（VOLTAIRE）

In love, when one person takes a knock, it is the other who sheds the tears.
对相爱的两个人而言，当一方遭受打击，落泪的会是另一方。

——斯图尔特与琳达·麦克法兰（STUART & LINDA MACFARLANE）

You are washed with the whitest fire of life – when you take a woman you love – and understand.

如果你爱她，懂她，那么你将承蒙最纯净的生命之火的洗礼。

——D.H. 劳伦斯（D. H. LAWRENCE）

We take our own magic with us you and I.
我们自带魔力，这个魔力就是你和我。

————帕姆·布朗（PAM BROWN）

Love is something sent from Heaven to worry the hell out of you.
爱情是天赐的礼物，让你六神无主。

——多莉·帕顿（DOLLY PARTON）

The most beautiful word in any language is our own name spoken to us with love by somebody we love.

在任何语言中，最美丽的词藻是听到我们所爱的人无限爱意地呼唤我们的名字。

——杰弗里·马森（JEFFREY MASSON）

Love comforteth, like sunshine after rain,
Love's gentle spring doth always fresh remain.

爱像雨后的阳光，给人慰藉；
爱如温和的春日，永葆清新。

——威廉·莎士比亚（WILLIAM SHAKESPEARE）

Whenever you share love with another, you'll notice the peace that comes to you and to them.

与他人分享爱的时刻，你就能感受到宁静降临到彼此身边。

——特蕾莎修女（MOTHER TERESA）

Keep love in your heart. The consciousness of loving and being loved brings a warmth and richness to life that nothing else can bring.

心中有爱。感受到爱与被爱，都将给你的生活带来无比的温暖与富足。

——奥斯卡·王尔德（OSCAR WILDE）

Date: / /

Mon.	Tue.	Wed.	Thu.	Fri.	Sat.	Sun.

There is a way from your heart to mine and my heart knows it.

我知道，有一条路从你的心通往我的心。

——札拉丁·鲁米（JALAL AL-DIN RUMI）

Love is not merely the indulgence of one's personal taste buds, it is also the delight in indulging another's.

爱情不仅仅是沉浸在自己的喜悦中，也要为别人的幸福由衷的快乐。

——劳里·李（LAURIE LEE）

What is important is that one is capable of love. It is perhaps the only glimpse we are permitted of eternity.

有能力去爱很重要，这爱的能力或许可以使我们永恒。

——海伦·海斯（HELEN HAYES）

Love is a settled heart and a big heart. Love encompasses everything inside of you.
爱是拥有一颗坚定、博大的心。爱包含了你内心的一切。

——斯图尔特·王尔德（STUART WILDE）

Date: / /

Mon.	Tue.	Wed.	Thu.	Fri.	Sat.	Sun.

The most important thing in life is to learn how to give out love, and to let it come in.
人生最重要的事情是学会如何爱与被爱。

——莫里·施瓦茨（MORRIE SCHWARTZ）

女郎，你的笑声中跳跃着生命之泉的音符。

——拉宾德拉纳特·泰戈尔（RABINDRANATH TAGORE）

Woman, in your laughter you have the music of the fountain of life.
女郎，你的笑声中跳跃着生命之泉的音符。
——拉宾德拉纳特·泰戈尔（RABINDRANATH TAGORE）

Love is the strongest of nature's forces – able to bring joy even out of tragedy.

爱是自然界最强大的力量，甚至能够从悲剧中带来快乐。

——斯图尔特与琳达·麦克法兰（STUART & LINDA MACFARLANE）

Treasure the love that you receive above all. It will survive long after your gold and good health have vanished.

珍惜你所得到的爱吧。当财富和健康都离你远去，还有爱会长存。

——奥格·曼迪诺（OG MANDINO）

Date: / /

Mon.	Tue.	Wed.	Thu.	Fri.	Sat.	Sun.

For a lot of us love is the central drama of our lives. It's the thing for which we take inconceivable risks and make moral decisions that we can't imagine ourselves making.

对我们许多人而言，爱情是生活的中心剧目。为了爱情，我们甘冒不可思议的风险，做出自己都无法想象的道德抉择。

——安妮塔·施里夫（ANITA SHREVE）

Date: /   /

| Mon. | Tue. | Wed. | Thu. | Fri. | Sat. | Sun. |

120

A marriage makes of two fractional lines a whole; it gives to two purposeless lives a work, and doubles the strength of each to perform it.

婚姻使两条分离的线段合二为一。它给予两个漫无目的的生命一项共同的任务，并且赋予婚姻各方双倍的力量去完成这项任务。

——马克·吐温（MARK TWAIN）

Oh, the miraculous energy that flows between two people who care enough to take the risks of responding with the whole heart.

啊，一股神奇的力量在两个相爱的人之间流动，这力量让他们不惧危险，全心投入，心心相印。

——亚历克斯·诺布尔（ALEX NOBLE）

Two things cannot alter,
Since Time was, nor today:
The flowing of water;
And Love's strange, sweet way.

有两件事亘古未变:
河水奔流不息;
爱情炽烈又甘甜。
　　——日本抒情诗（JAPANESE LYRIC）

123

Date: / /

| Mon. | Tue. | Wed. | Thu. | Fri. | Sat. | Sun. |

After him I love more than I love these eyes, more than my life, more, by all mores...

我爱他，胜过爱自己的双眸，胜过爱自己的生命，胜过爱一切……

——威廉·莎士比亚（WILLIAM SHAKESPEARE）

Knowing is the most profound kind of love, giving someone the gift of knowledge about yourself.
懂得是最深沉的爱，这会让他更理解你。

——玛莎·诺曼（MARSHA NORMAN）

Breathless, we flung us on the windy hill,
Laughed in the sun, and kissed the lovely grass.

我们在微风和煦的山坡上奔跑着，气喘吁吁；

我们在阳光下欢笑着，亲吻着茵茵草地。

——鲁伯特·布鲁克（RUPERT BROOKE）

Love is the only light that can truly read the secret signature of the other person's individuality and soul.

唯有在爱的照耀下，才能真正理解他人个性与灵魂的独特魅力。

——约翰·奥多诺休（JOHN O'DONOHUE）

Date: / /

Mon.	Tue.	Wed.	Thu.	Fri.	Sat.	Sun.

Days are more precious waking up with someone you love.
与爱人一同醒来的日子更加可贵。

——卡莉·西蒙（CARLY SIMON）

The heart has its reasons which reason knows nothing of.
心有所向，而理智却对此一无所知。

——布莱斯·帕斯卡尔（BLAISE PASCAL）

There is only one element in life which is worth having at any cost, and it is love. Love immense and infinite, broad as the sky and deep as the ocean – this is the one great gain in life. Blessed is he who gets it.

生命中唯一值得倾其所有去拥有的东西是爱情。爱情宽广无边，如天空般浩瀚，如大海般深邃——爱情是生命中最大的收获，获得爱情的人是何等的幸运啊。

——斯瓦米·维韦卡南达（SWAMI VIVEKANANDA）

Mon.	Tue.	Wed.	Thu.	Fri.	Sat.	Sun.

If we are lacking in love, real love – the kind that spends itself unreservedly on another – we accomplish nothing.

假如生活中缺少爱，没有毫无保留付出的真爱，生活该是多么苍白啊！

——新斯凯特僧团（THE MONKS OF NEW SKETE）

But there's nothing half so sweet in life as love's young dream.
人生中最甘甜的部分莫过于年少时的爱情美梦。

——托马斯·莫尔（THOMAS MOORE）

I do not think we can ever adequately define or understand love; I do not think we were ever meant to. We are meant to participate in love without really comprehending it. We are meant to live into love's mystery.

我认为我们永远无法充分定义或者理解爱情，我认为我们从未认真思量过爱情。我们注定要在不懂爱的时候去爱恋，我们注定要活在爱情的神秘之中。

——杰拉尔德·G. 梅（GERALD G. MAY）

Date: / /

Mon.	Tue.	Wed.	Thu.	Fri.	Sat.	Sun.

Truly loving another means letting go of all expectations.
真爱一个人意味着放下所有的期待。

——卡伦·凯斯（KAREN CASEY）

Happiness, when love adds one to one, is not doubled but multiplied a thousand times.

两情相悦时，幸福可不是简单叠加，而是被放大一千倍。

——斯图尔特与琳达·麦克法兰（STUART & LINDA MACFARLANE）

Date: /   /

Mon.	Tue.	Wed.	Thu.	Fri.	Sat.	Sun.

You have intensified all colours, heightened all beauty, deepened all delight. I love you more than life, my beauty, my wonder.

你让所有的色彩更加绚丽，让所有的美丽更加夺目，让所有的欢乐更加深刻。
我爱你胜过爱自己的生命，我的佳人，我的奇迹。

——艾尔弗雷德·达夫·科珀爵士（SIR ALFRED DUFF COOPER）

Friendship and love should be the safe regions where your unknown selves can come out to play.

友谊和爱情是放飞未知自我最安全的花园。

——约翰·奥多诺休（JOHN O'DONOHUE）

To experience love in ourselves and others, is the meaning of life.
用心感受，爱自己，爱他人，这就是人生的意义。

——玛丽安·威廉森（MARIANNE WILLIAMSON）

When love floods the senses, it jams your sonar, blinds you to all else. Lightning might crash arou
you, eels and piranhas nibble at your toes and you don't care, because only One Thing matters – t
longing which has overtaken your soul. We humans were made for that sweet, sweeping sickness.

沉浸在爱情里时，你会忽视周围的一切。雷电可能就在你身边轰鸣，鳗鱼和食人鱼可能
你的脚趾，而你并不在意。因为在你心里，你只在乎一件事——对爱的渴望已超越你的
魂，我们人类生来就乐意患上这种甜蜜而又深沉的病痛。

——丝·蒙哥马利（SY MONTGOMERY

Date: /   /

Mon.	Tue.	Wed.	Thu.	Fri.	Sat.	Sun.

Each year that's gone has brought us new kinds of happiness – may every year to come bring even greater joy.

过去的每一年都带给我们新的幸福——愿我们以后一年比一年快乐。

——帕姆·布朗（PAM BROWN）

Love is a plant of tenderest growth: treat it well, take thought for it and it may grow strong and perfume your whole life.

爱情是最娇嫩的植物。善待它吧，呵护它吧，它会茁壮成长，让您一生芬芳。

——弗兰克·哈里斯（FRANK HARRIS）

Love begets life. Life begets love.
爱情滋养生活，生活孕育爱情。
——斯图尔特与琳达·麦克法兰（STUART & LINDA MACFARLANE）

Love is everything it's cracked upto be. That's why people are so cynical about it…. It really is worth fighting for, being brave for, risking everything for. And the trouble is, if you don't risk anything, you risk even more.

爱情易碎，所以人们常常质疑爱情。但爱情值得人们为之奋斗，鼓足勇气，为之冒险。而我们的烦恼在于，如果你拒绝为爱情冒风险，那么你将来会承担更大的风险。

——埃丽卡·容（ERICA JONG）

Accustom yourself to continually make many acts of love, for they enkindle and melt the soul.

让自己习惯于做好事善事吧，因为它们能点亮并且感化灵魂。

——圣特蕾莎修女（SAINT TERESA）

An orange on the table, your dress on the rug, and you in my bed, sweet present of the present, cool of night, warmth of my life.

夜色清凉，而你温暖了我的生活。桌上有橘子，地毯上堆着你的衣服，你我相拥而眠。这甜蜜的时刻，便是生活的温暖。

——雅克·普莱维尔（JACQUES PRÉVERT）

Date:　　/　　/

Mon.	Tue.	Wed.	Thu.	Fri.	Sat.	Sun.

How glorious it is to live within your love.
沐浴在你的爱情里是多么美好绚烂。

——帕姆·布朗（PAM BROWN）

...it is in loving, as well as in being loved, that we become most truly ourselves. No matter what we do, say, accomplish, or become, it is our capacity to love that ultimately defines us. In the end, nothing we do or say in this lifetime will matter as much as the way we have loved one another.

……我们只有在爱与被爱的过程中，才能成为最真实的自己。无论我们做了什么，说了什么，成就了什么，或者成为了什么人，最终只有我们的爱能来刻画我们自己。终其一生，我们的言行都不如我们彼此相爱的方式重要。

——达夫妮·罗丝·金马（DAPHNE ROSE KINGMA）

love......widens and enriches our life.
爱，拓宽并丰富我们的生活。

——弗兰茨·卡夫卡（FRANZ KAFKA）

Date: / /

Mon. Tue. Wed. Thu. Fri. Sat. Sun.

The one thing we can never get enough of is love.
And the one thing we never give enough of is love.
我们永不满足的是爱情，而我们永远无法充分给予的也是爱情。

——亨利·米勒（HENRY MILLER）

In sandy earth or deep in valley soil I grow, a wildflower thriving on your love.
我是生长在沙土里或山谷深处的一朵野花，沐浴着你的爱情茁壮成长。
——摘自《圣经·雅歌》（FROM "SONG OF SONGS"）

How quickly bodies come to love each other, promise themselves to each other, without asking permission from the mind!

无须理智的允许，身体之间就可以如此迅速地彼此承诺、彼此相爱！

——罗莉·穆尔（LORRIE MOORE）

One of the most important things is to have love in your life; to give love and hopefully to be able to accept it.

人生中最重要的事情就是拥有爱，爱他人，也被人爱。

——伊娃·弗雷泽（EVA FRASER）

Love is what it is: the most complicated, intense and indefinable emotion.
And yet without it…well, life wouldn't really be worth living, would it?
爱情应该是它本来的样子：最复杂、最强烈、最难以形容的情感。如
果没有爱情……活着还有什么意义呢？

——麦克·盖尔（MIKE GAYLE）

Date: /   /

Mon.	Tue.	Wed.	Thu.	Fri.	Sat.	Sun.

The act in bed is the mainstay of creation, and it may satisfy the senses but it does not feed the heart. Love needs the instrument of a voice.

床第之欢是造物主的恩赐，它带给我们感官的满足，却无法抚慰心灵。而爱情需要听到心灵的歌唱。

——凯瑟琳·库克森（CATHERINE COOKSON）

How comfortable we are together, safe from the world in mutual trust, in love and friendship.

我们俩在一块儿，多么舒心，多么惬意，远离尘嚣，彼此信任，沐浴着爱情和友谊。

——帕姆·布朗（PAM BROWN）

Date: / /

Mon. Tue. Wed. Thu. Fri. Sat. Sun.

Is there anything on earth more unearthly than to be in love at eighteen? It is like an abundant spring garden. My heart was the Orient, and the sun rose from it; I could have picked the stars from the sky.

世上还有什么比十八岁的恋爱更神秘？它仿佛繁花似锦的春日花园。我的心儿就是旭日东升的起点，我快乐得仿佛可以摘到天上的星星。

——让·斯塔福德（JEAN STAFFORD）

Young love. It is the greatest thing.
年轻人的爱情，是最伟大的事情。

——多萝西·帕克（DOROTHY PARKER）

Date:　　/　　/

Mon.	Tue.	Wed.	Thu.	Fri.	Sat.	Sun.

You trust the man you love with every single part of you …heart, body, soul, mind, life, bank balance, worries, fears, secrets, hopes and dreams.

用你的心灵、身体、灵魂、思想、生活、银行存款、担心、恐惧、秘密、希望以及梦想……用你的一切去相信你所爱的人。

——希安·E. 摩根（SIÂN E. MORGAN）

As we are, here, together, now and here,
Always you and I.
我将永远和你在一起，就像我们此时此地在一起。
——罗伯特·格雷夫斯（ROBERT GRAVES）

...everything you do souses me, terrifies me, tortures me, elates me, everything you do is perfect.

……你做任何事都令我牵挂，令我伤感，有时我感到恐惧，有时我备受煎熬，我又时常兴奋不已，你做的一切都完美无暇。

——保罗·艾吕雅（PAUL ÉLUARD）

The heart that loves is always young.
心中有爱，永远年轻。

——希腊谚语（GREEK PROVERB）

Nothing is more mysterious, mystical or magical than falling in love.
没有什么事能比坠入爱河更神秘、更有魔力、更神奇。
——萨拉·班恩·布瑞斯纳克（SARAH BAN BREATHNACH）

Love is the strength that binds people together. It is the most powerful thing of all.
是爱把人们紧紧联系在一起，爱是最强大的力量。
——比利·米尔斯（BILLY MILLS）

Love is enriched by every good thing shared – and made stronger by every sorrow faced together.

分享每一次美好的经历，能使爱情更丰盈；而共同经历每一次悲伤，能使爱情更坚强。

——帕姆·布朗（PAM BROWN）

I am at one with you. When we are together, I am content and relaxed in a way that I have never felt with any other person. Being with you is like being wrapped up in a warm, comfortable blanket of love.

和你在一起时，我找到了自己，我变得完整。只有和你在一起，我才感到前所未有的满足与放松。和你在一起，犹如裹在一张温暖舒适、爱意浓浓的毯子里。

——斯图尔特与琳达·麦克法兰（STUART & LINDA MACFARLANE）

I tell you, the more I think, the more I feel that there is nothing more truly artistic than to love people.

让我告诉你吧，我越想越觉得没有什么事能比爱一个人更具有艺术魅力了。

——文森特·梵高（VINCENT VAN GOGH）

The only thing I know about love is that love is all there is....
我对爱唯一的认知就是：爱就是一切……

——埃米莉·狄更生（EMILY DICKINSON）

There were moments when I feared to hear your voice, and then I was disconsolate that it was not your voice.
So many contradictions, so many contrary movements are true, and can be explained in three words: I LOVE YOU.

有时，我害怕听到你的声音，但又会因为那不是你的声音而感到难过。
我是如此的矛盾啊，而这一切又是那样的真实，我可以用三个字解释：我爱你。

——朱莉·德·莱斯皮纳斯（JULIE DE L'ESPINASSE）

Love is in ordinary things – small kindnesses, a hand on the shoulder, a kiss in passing. A dish of strawberries.

不经意的温柔之举，轻搭在肩上的一只小手，走过身边时的一个吻，一盘草莓——平凡之中蕴藏着爱情。

——夏洛特·格雷（CHARLOTTE GRAY）

Life is such a bizarre ball game, the only thing to hold on to is this stupid thing called love.

生活就像一场奇妙的球赛，唯一让人们可以坚持到底的力量源自这个被称为"爱"的愚蠢玩意儿。

——安德鲁·邓肯（ANDREW DUNCAN）

Love is an irresistible desire to be irresistibly desired.
爱情是不可抗拒、无法抑制的渴望。

——罗伯特·弗罗斯特（ROBERT FROST）

When one has once fully entered the realm of love, the world – no matter how imperfect – becomes rich and beautiful, for it consists solely of opportunities for love.

一旦你完全走进爱情的王国，那么这个世界 ——无论多么不完美——都会变得丰富而美丽，因为这里只给爱留有机会。

——索伦·克尔凯郭尔（SØREN KIERKEGAARD）

First of all things, there must be that delightful, indefinable state called feeling at ease with your companion, the one man, the one woman out of a multitude who interests you, who meets your thoughts and tastes.

最重要的是，当你与爱人相伴时，一定会感到一种让你愉悦、无法形容的自在。在你关注的人群之中，唯有他（她）既与你心灵相通，又使你称心如意。

——朱莉亚·杜林（JULIA DUHRING）

Lay in my arms till break of day then tarry a little while longer.
躺在我的臂弯，直至天亮，然后再多停留片刻吧。

——琳达·麦克法兰（LINDA MACFARLANE）

You are always new. The last of your kisses was ever the sweetest; the last smile the brightest.

你日日如新。你刚才那一吻，最甜蜜；刚才那微笑，最灿烂。

——约翰·济慈（JOHN KEATS）

Love isn't decent. Love is glorious and shameless.
爱情不是得体礼貌。爱情灿烂夺目，令人手足无措。
——伊丽莎白·冯·阿尼姆（ELIZABETH VON ARNIM）

There is nothing you can do, achieve or buy that will outshine the peace, joy and happiness of being in communion with the partner you love.

你的能力、你的成就、你的财富，都无法和与你所爱之人交流时的平和、快乐和幸福媲美。

——德尔斯·伊夫琳与保罗·莫斯奇塔（DRS EVELYN AND PAUL MOSCHETTA）

How can I count the kindness, the astonishments, the joys that you have given me? Your strength, your laughter, the comfort of your arms.

我怎样才能细数你赐予我的美好、惊奇和欢愉？你的力量、你的欢笑和你臂弯的温暖。

——帕姆·布朗（PAM BROWN）

Time flies, suns rise, and shadows fall – Let them go by, for love is over all.
时光荏苒，日出日落——让这些都逝去吧，因为爱胜过一切。
　　　　　　　——发现于某日晷上（FOUND ON A SUNDIAL）

Love heals, love is what makes things a little better than before. Love is Universal.

爱能治愈伤痕，爱使一切变得更加美好。爱无处不在。

——熊心（BEARHEART）

Mon.	Tue.	Wed.	Thu.	Fri.	Sat.	Sun.

Shall I compare thee to a summer's day?
Thou art more lovely and more temperate.
我怎么能够把你来比作夏天?
你不独比它可爱也比它温婉。

——威廉·莎士比亚（WILLIAM SHAKESPEARE）

We do amazing things for love: take risks, conquer our fears and our limiting beliefs – all because of love. It can truly create magic in our lives....

我们为了爱情敢于做出许多奇妙的事情：冒险、克服内心的恐惧以及狭隘的执念——所有这一切皆是为了爱情。爱情确实能够在我们的人生中创造奇迹……

——理查德·帕克斯·科多克（RICHARD PARKES CORDOCK）

If I were pressed to say why I love him, I feel that my only reply could be: "Because it was he, because it was I".

如果非要我解释为什么会爱他，我觉得唯一的回答就是："因为是他，因为是我。"

——米歇尔·德·蒙田（MICHEL EYQUEM DE MONTAIGNE）

To live is to love. To love is to live.
活着就要去爱，爱就是活着。
——斯图尔特与琳达·麦克法兰（STUART & LINDA MACFARLANE）

Without you, dearest dearest I couldn't see or hear or feel or think – or live – I love you so and I'm never in all our lives going to let us be apart another night. It's like growing old, without you. I want to kiss you so – I love you – and I can't tell you how much.

如果没有你，我最亲爱的，我就无法看见，无法倾听，无法感知，也无法思考——甚至，无法活下去。我是如此爱你，在我们的一生之中，我永远不会让我们再分开哪怕一晚。如果没有你，将如同迟暮垂老。我多么想亲吻你——我爱你——我无法形容这份爱有多么热烈。

——塞尔达·菲茨杰拉德（ZELDA FITZGERALD）

Love's not Time's fool, though rosy lips and cheeks
Within his bending sickle's compass come;
Love alters not, with his brief hours and weeks,
But bears it out even to the edge of doom.

爱情无畏时间，
即使岁月的镰刀掠过，红颜不再。
爱情不因瞬息变化而改变，
它会屹立至命运的尽头。

——威廉·莎士比亚（WILLIAM SHAKESPEARE）

Love attracts love.
心有灵犀，两情相悦。

——阿娜伊斯·尼恩（ANAÏS NIN）

Date: / /

Mon.	Tue.	Wed.	Thu.	Fri.	Sat.	Sun.

Certainly I am happy when the music lifts the corner of your mouth and you smile. Your hand is very warm in mine. And you love me.

当音乐在你嘴角扬起时，当你微笑时，我是多么幸福啊。我握着你的手，暖意融融，并且你爱着我。

——帕丁·克莱（PADDI CLAY）

I did not know what love was till I met you. You taught me passion, showed me romance. And now I know life would be worthless without you.

直到遇见你，我才知道什么是爱。你教会我激情，赐予我浪漫。如今，我深知，没有你，人生将毫无价值。

——琳达·麦克法兰（LINDA MACFARLANE）

We think we love someone for their looks; their walk, maybe tone of voice, touch –
but when you analyse it, it is really for their qualities – their warmth, their humour,
their intelligence, kindness, etc....

我们以为，我们会因为相貌、走路的姿态，也许还会因为说话的语调或者个性而
爱一个人——但是，经过分析，你会发现，使我们真正动心的是一个人的热情、
幽默、睿智和善良等品质。

——乔伊斯·格伦费尔（JOYCE GRENFELL）

In the arithmetic of love, one plus one equals everything, and two minus one equals nothing.

在爱情的算术中，一加一等于一切，而二减一等于零。

——米尼翁·麦克劳克林（MIGNON MCLAUGHLIN）

Your voice is the pillow on which I rest my heart, the blanket with which I warm my dreams.

你的声音是我心休憩的甜枕，是温暖我梦想的绒毯。

——查尔斯·吉格纳（CHARLES GHIGNA）

Love cannot be defined in one single term,
It cannot be taught and cannot be measured.
It should always be handled carefully,
An eternal and precious gift to always be treasured.

爱情无法只用一个词来定义，
它不需要训练，也无法度量。
爱情应该始终被小心呵护，
它是一份永恒且珍贵的礼物，将永
远被珍惜。

——雷迪·福克斯（REDDY FOX）

Date:　　/　　/

Mon.	Tue.	Wed.	Thu.	Fri.	Sat.	Sun.

What is love? Not to wish to exchange a hut for a palace.

什么是爱情？纵然千金华宅，不如我的茅屋爱巢。

——谚语（PROVERB）

Oh, the comfort – the inexpressible comfort, of feeling safe with a person – having neither to weigh thoughts nor measure words, but pouring them out.

啊，那种惬意——无以言表的惬意，是与心爱的人儿在一起时的无忧无虑——不必深思熟虑，也无须斟言酌句，只需倾吐真心、互诉衷肠。

——黛娜·马洛克·克雷克（DINAH MULOCK CRAIK）

All pleasures should be taken in great leisure and are worth going into in detail; love is not like eating a quick lunch with one's hat on.

爱情可不是连帽子都不摘匆匆吃个快餐，所有的乐趣都应该在最悠闲的时光里细细品味，享受欢乐。

——梅·韦斯特（MAE WEST）

...I have never known any union so sweet, and beautiful, so spiritual and soul-satisfying as ours. I swell in sunshine and my heart sings with happiness.

……我未曾见过像我们这般甜蜜、美丽的结合，精神与灵魂都得以满足。我沐浴在阳光里，心儿在幸福地歌唱。

——莱拉·塞科尔（LELLA SECOR）

Date: /   /

Mon.	Tue.	Wed.	Thu.	Fri.	Sat.	Sun.

Lovers re-create the world.
这个世界是由相爱的人重新创造的。

——卡特·海沃德（CARTER HEYWARD）

May we find strength that comes from unity, the quiet joy that comes from long companionship.
愿我们从相依相偎中获得力量，从长情陪伴中收获怡人的欢乐。

——帕姆·布朗（PAM BROWN）

But love is a durable fire in the mind ever burning; Never sick, never old, never dead from itself never turning.

爱情是心中永远燃烧的火焰。爱情不会生病，不会衰老，不会死亡，永远不褪色。

——沃尔特·罗里（WALTER RALEIGH）

Come live with me and be my love.
走进我的生活，做我的爱人吧。

——克里斯托弗·马洛（CHRISTOPHER MARLOWE）

There's only one thing in this world that's worth having. Love. L-o-v-e. You love somebody, somebody loves you. That's all there is to it.

世上唯有一事值得拥有，那就是爱。你爱他，他也爱你。这就是全部。

——查尔斯·梅吉恩达尔（CHARLES MERGENDAHL）

It is good to love as many things as one can, for therein lies true strength, and those who love much, do much and accomplish much, and whatever is done with love is done well. Love is the best and the noblest thing in the human heart, especially when it is tested by life.

尽你所能去爱一切，是件美好的事情，因为爱中蕴藏着真正的力量。那些爱得更多、做得更多的人，也会有更多成就。心怀爱意做事，皆有善终。爱是人们心中最美好、最高尚的东西，经受过生活考验的爱尤其美好、尤其高尚。

——文森特·梵高（VINCENT VAN GOGH）

I need your love as a touchstone of my existence. It is the sun which breathes life into me.

我需要你的爱作为我存活于世的证明，你的爱就是给予我生命的太阳。

——朱丽叶·杜洛埃（JULIETTE DROUET）

Nothing can be greater than love. Love is life, and life itself is spontaneous nectar and delight.

没有什么会比爱更伟大。爱就是生命，而生命本身自然就会产生甘露与欢乐。

——斯里·钦莫伊（SRI CHINMOY）

Date: / /

Mon.	Tue.	Wed.	Thu.	Fri.	Sat.	Sun.

When two people are at one in their inmost hearts, they shatter even the strength of iron or bronze. And when two people understand each other in their inmost hearts, their words are sweet and strong, like the fragrance of orchids.

二人同心，其利断金；同心之言，其臭如兰。

——《易经》（THE BOOK OF CHANGES）

True love has no sides, limits, or corners. It is without circumference and beyond inside and out. The heart of limitless love includes all and everything, embracing one and all in its warmth. Genuine love is enough in simply being itself.

真爱无边、无限、无角落，真爱浩瀚无边。一颗充满无限爱意的心包容一切、用温暖拥抱一切，真爱本身就已足够。

——舒亚·达斯喇嘛（LAMA SURYA DAS）

It is more fun watching someone you love have fun than trying to have fun yourself.

看到你爱的人开心，比努力让自己开心更令人愉悦。

——蕾切尔·约翰逊（RACHEL JOHNSON）

There is no remedy for love but to love more.
唯有加倍去爱，除此之外，别无灵丹妙药。
——亨利·戴维·梭罗（HENRY DAVID THOREAU）

209

Mon.	Tue.	Wed.	Thu.	Fri.	Sat.	Sun.

If I have the colic,
I take some medicine.
If I am seized by the pox,
I go down to the Hot-Springs.
But where is there help,
for what SHE does to me?

如果我有心绞痛,
我可以吃药;
如果我出水痘,
我会去泡温泉;
但是, 如果她闯进了我的心,
我该何处求救?

——埃塞俄比亚爱情诗 (ETHIOPIAN LOVE POEM)

Your heart is an inexhaustible spring, you let me drink deep, it floods me, penetrates me, I drown.

你的心是永不干涸的甘霖，请让我尽情畅饮吧。让你的甘露浸润我，穿透我，我要沉醉在你的心田。

——古斯塔夫·福楼拜（GUSTAVE FLAUBERT）

What else is the world interested in? What else do we all want, each one of us, except to love and be loved, in our families, in our work, in all our relationships?

在家庭中，在工作中，在所有的关系中，世人都是在寻求爱与被爱，除此之外，别无所求。

——多萝西·戴（DOROTHY DAY）

Thou art to me a delicious torment.
对我而言，你是一种美妙的折磨。

——拉尔夫·沃尔多·爱默生（RALPH WALDO EMERSON）

Date:　　/　　/

Mon.	Tue.	Wed.	Thu.	Fri.	Sat.	Sun.

The greatest beauty secret of all is love. To love and to be loved. A woman in love has a certain glow.

最伟大的美丽秘诀是爱情，爱与被爱，恋爱中的女人最美丽。

——琼·科林斯（JOAN COLLINS）

I enjoy nothing without you. You are the prism through which the sunshine, the green landscape, and life itself, appear to me.... I need your kisses upon my lips, your love in my soul.

没有你，我百无聊赖。你就是那面多棱镜，透过你，我看到了阳光、绿色的风景和生命……我要你吻我的唇，在我灵魂中留下你爱的印迹。

——朱丽叶·杜洛埃（JULIETTE DROUET）

All that we love deeply becomes a part of us.
我们深爱的一切终会成为我们的一部分。

——海伦·凯勒（HELEN KELLER）

We are bound by love to share all things, to explore the world together, to learn from each other, to discover compromise, and patience and a growing, deepening love.

我们因爱结合, 我们分享一切, 共同探索世界, 相互学习; 我们学会妥协, 学会忍耐并且发现日渐深厚的爱。

——玛丽昂· C. 加瑞缇 (MARION C. GARRETTY)

...love means that I am confident enough about that other that I can trust him with my gift.

……爱情意味着我对另一个人有足够的信心，我天然地信任他。

——无名氏（AUTHOR UNKNOWN）

We do not know what excitements or adventures, what dangers or amazements lie ahead, or where the roadway leads. But simply to travel in each other's company is happiness and certainty enough.

我们并不知道前方会有什么激动人心的经历或冒险，会有什么危险或惊喜，也不知道这条路通往何方。但是，彼此结伴旅行就已经足够幸福、足够肯定。

——帕姆·布朗（PAM BROWN）

We are crazy. People have said it. We know it. Yet we go on. But being crazy together is just fine.

大家都说我们疯了，我们的确疯了，我们还要继续疯狂，我们能够一起疯狂真是太棒了。

——雷·布雷德伯里（RAY BRADBURY）

I love you as one must love: excessively, to the point of madness and despair.
There are two things which must never be mediocre: poetry and love....

我爱你，因为我必须爱你：爱到极致、爱到疯狂、爱到绝望。有两件事永远
不会平庸：诗歌和爱情……

——朱莉·德·莱斯皮纳斯（JULIE DE L' ESPINASSE）

Date:　　/　　/

Mon.	Tue.	Wed.	Thu.	Fri.	Sat.	Sun.

Love – bittersweet, irrepressible – loosens my limbs and I tremble.
爱情——痛苦又甜蜜，我们无法压抑爱情，爱情释放了我的天性，让我浑身颤栗。

——莎孚（SAPPHO）

Nearly every one of us is starving to be appreciated, to be the recipient of that most supreme compliment – that we are loved.

人人都渴望得到别人的欣赏，渴望得到最由衷的赞扬，我们渴望被人爱着。

——利奥·巴斯卡利亚（LEO BUSCAGLIA）

Date: / /

Mon.	Tue.	Wed.	Thu.	Fri.	Sat.	Sun.

如若无人疼爱我们，我们也就不会再爱自己了。

We cease loving ourselves if no one loves us.
如若无人疼爱我们，我们也就不会再爱自己了。
——斯达尔夫人（MADAME DE STAEL）

To love someone means to be involved with, to identify with, to engage with, to suffer with and for them, and to share their joys.

爱一个人意味着融入他的生活。认同他，适应他，支持他，与他同甘共苦。

——威利亚德·盖林（WILLARD GAYLIN）

True love speaks in tender tones and hears with gentle ear, true love gives with open heart and true love makes no harsh demands. It neither rules nor binds.

真爱是柔声细语、用心倾听，真爱是敞开心扉去给予。真爱不是苛求，真爱不是控制，也不是束缚。

——无名氏（AUTHOR UNKNOWN）

I know your eccentricities, your moods. And somehow, for some reason I can never fully understand, I am crazy with love for you.

我了解你的怪癖、你的情绪。但不知道为什么，我永远无法完全了解自己为何疯狂地爱你。

——夏洛特·格雷（CHARLOTTE GRAY）

Date:　　/　　/

Mon.	Tue.	Wed.	Thu.	Fri.	Sat.	Sun.

The course of true love never did run smooth.
真爱的路途上布满荆棘。

——威廉·莎士比亚（WILLIAM SHAKESPEARE）

You are everything I need. You are the sun, the air I breathe. Without you, life wouldn't be the same. Please never go away. And if you go, then don't forget to take me with you.

你是我所需要的一切。你是我的太阳，是我呼吸的空气。如果没有你，我的生活将了无生趣。请永远不要离开我，如果你要走，别忘了，把我带上，与你同行。

——巴士雅（BASIA）

We must love one another, yes, yes, that's all true enough, but nothing says we have to like each other.

我们必须相爱，是的，是的，千真万确，但我们不必彼此喜欢。

——彼得·德·弗里斯（PETER DE VRIES）

Where thou art, there is the world itself.
你在哪里，哪里就是整个世界。

——威廉·莎士比亚（WILLIAM SHAKESPEARE）

Hold tenderly that which you cherish.
温柔地拥抱你所珍惜的一切。

——鲍勃·艾伯蒂（BOB ALBERTI）

Darling, you want to know what I want of you. Many things of course but chiefly these, I want this thing we have inviolate and waiting – the person who is neither I nor you but us.

亲爱的，你想知道我希望从你那里得到什么。我当然想要许多，但是，最重要的是，我希望我们拥有的这件东西坚如磐石，也是我一直期待的，那就是，我希望我们彼此不再是你，是我，而是我们。

——约翰·斯坦贝克（JOHN STEINBECK）

When a man loves a woman and that woman loves him, the angels leave heaven and come to their house and sing.

当一个男人爱一个女人，而这个女人也爱他，天使们就会从天堂来到他们的爱巢，为他们歌唱。

——布拉马·库玛利斯（BRAHMA KUMARIS）

A happy couple share a certain smile that no one else quite understands.
幸福爱侣之间的默契一笑，旁人难以体会。

——帕姆·布朗（PAM BROWN）

In dreams and in love there are no impossibilities.
在梦想与爱情中，一切皆有可能。

——贾纳斯·阿罗尼（JANUS ARONY）

Love is something eternal – the aspect may change, but not the essence.

爱情不朽——外表可能改变，但芬芳永存。

——文森特·梵高（VINCENT VAN GOGH）

There is only one happiness in life, to love and be loved.
生命中只有一种幸福：爱与被爱。

——乔治·桑德（GEORGE SAND）

If I had a single flower for every time I think about you, I could walk forever in my garden.

假如每次想起你，我都会得到一朵鲜花，那么，我可以永远在我的花园里徜徉。

——黛安娜·阿克曼（DIANE ACKERMAN）

Date: / /

Mon.	Tue.	Wed.	Thu.	Fri.	Sat.	Sun.

Love… makes one little room, an everywhere.
是爱将平凡的小屋打造成了梦想家园。

——约翰·多恩（JOHN DONNE）

I love you for the part of me that you bring out.
我爱你，因为你让我活出了真实的自己。

——罗伊·克罗夫特（ROY CROFT）

All the goals and targets in the world mean nothing unless you're happy, you love and you're loved.

除非你是快乐的，你爱着并且被爱着，否则，世界上的一切目的和目标都毫无意义。

——达赖厄斯·达拉斯（DARIUS DANESH）

As long as one can admire and love, then one is young forever.

只要你还能去欣赏、去爱，你就永远年轻。

——帕布罗·卡萨尔斯（PABLO CASALS）

How silver-sweet sound lovers' tongues by night,
Like softest music to attending ears!
夜里，恋人的柔声细语甜美如银铃，
仿佛最温柔悦耳的音乐！

——威廉·莎士比亚（WILLIAM SHAKESPEARE）

In you alone my desires give birth to delirium, in you alone my love bathes in love.

唯有你，能让我肆意奢望；唯有你，能让我沐浴爱河。

——保罗·艾吕雅（PAUL ÉLUARD）

Everybody's the same when it comes to love....When someone in the ghetto falls in love she hears bells – the same bells someone uptown hears when she falls in love.

在爱情面前人人都一样……无论高贵还是卑微，只要坠入爱河，都会听到同样的爱情钟声。

——贝里·戈迪（BERRY GORDY）

Love is when you can spend a day together doing nothing in particular –
and be supremely happy.
爱情就是两人无所事事地消磨一天，也会感到极度的快乐。

——帕姆·布朗（PAM BROWN）

A woman who is loved always has success.
被爱的女人，总是成功的。

——薇姬·鲍姆（VICKI BAUM）

Two lovely berries moulded on one stem:
So, with two seeming bodies, but one heart.

一对可爱的浆果共结一枝：看似分身，却是同心。

——威廉·莎士比亚（WILLIAM SHAKESPEARE）

The greatest waste one can leave behind in life is the love that has not been given.
人生最大的浪费是从未给予爱情。

——无名氏（AUTHOR UNKNOWN）

If love does not know how to give and take without restrictions or expectations, then it is not love, but a transaction.
没有约束，也无所期待，不懂得如何给予和索取的爱情，那么，这不是爱情，而是交易。

——埃玛·戈德曼（EMMA GOLDMAN）

There is nothing more precious in this world than the feeling of being wanted.

在这个世界上，被需要的感觉最为珍贵。

——黛安娜·多丝（DIANA DORS）

Unconditional love is the most precious gift we can give. Being forgiven for the past is the most precious gift we can receive.

我们所能给予的最珍贵的礼物是无条件的爱，我们所能收到的最珍贵的礼物是宽恕过去。

——萨拉·J.沃格特（SARAH J. VOGT）

I love you when you laugh so much you slither off the chair. When you sleep all sprawled and limp like a little child.

当你大笑着从椅子上滑下来时，我爱你；当你像小孩一样伸开四肢酣睡时，我也爱你。

——帕姆·布朗（PAM BROWN）

One look and my soul sings, one word and my heart flutters. One touch and my whole body thrills with delight.

你的一个眼神，就足以唤起我灵魂歌唱；你的只言片语，就足以让我小鹿乱撞；你轻柔的爱抚，就会让我幸福得浑身颤栗。

——琳达·麦克法兰（LINDA MACFARLANE）

I bless you. I kiss and caress every tenderly beloved place and gaze into your deep, sweet eyes which long ago conquered me completely.

我祝福你。我亲吻并爱抚着每一处曾被温柔爱过的地方，凝望着你深邃、甜美的双眸，你的这双眼眸早已把我彻底征服。

——扎瑞斯特·亚历山德拉（TSARITSA ALEXANDRA）

Love, I've come to understand, is more than three words mumbled before bedtime. Love is sustained by action, a pattern of devotion in the things we do for each other every day.

我终于明白，爱情，不止是睡前呢喃的那三个字。爱情需要行动来维系，它是我们每天为彼此奉献的一种方式。

——尼古拉斯·斯帕克思（NICHOLAS SPARKS）

Date:　　/　　/

Mon.	Tue.	Wed.	Thu.	Fri.	Sat.	Sun.

Thank you for being here. Thank you for everying.
谢谢你伴我左右，谢谢你所做的一切。

——马娅·帕特尔（MAYA PATEL）

Wherever I am, My heart is with you, my love. The river can not keep me from you.
In my mind's eye, I see always you, my love. Nothing can divide us, one from one.
My heart sings for you, my only love.
我的爱人，无论我身处何地，我心与你同在，河流无法把我们分开。
在我的脑海里，我看到的永远是你，我的爱人。没有什么能把我们分开。
我的心儿为你歌唱，我唯一的爱人。

——祖鲁人的爱情诗（ZULU LOVE POEM）

Loving one another is our only reason for being.
相亲相爱是我们存在的唯一理由。

——多莉·帕顿（DOLLY PARTON）

I have found comfort in your arms. Courage in weakness. Hope in despair.
Laughter Love.

在你的臂弯里，我找到了慰藉，于软弱中找回勇气，于绝望中看到希望，
在爱情中欢颜。

——帕姆·布朗（PAM BROWN）

Once again last night you would not let me sleep. Before I went to sleep I moved over and made room for you and tried to imagine you there so soft and warm and smooth. I put out a hand and was disappointed.

昨晚，你再次让我无法入眠。入睡前，我在床上为你腾出地方，想象你此刻就在我身旁，如此温柔、如此温暖、如此光滑。我伸出手去，却满是失望。

——鲍勃·格拉夫顿（BOB GRAFTON）

Don't you think I was made for you? I feel like you had me ordered – and I was delivered to you – to be worn – I want you to wear me, like a watch-charm or a buttonhole bouquet – to the world.

你不觉得我是为你而生的吗？我觉得你就是我的主宰，我是被送到你身边的，我希望你把我当作怀表上的表链或者别在钮扣眼里的小花束，把我戴在你身上，走遍世界。

——塞尔达·菲茨杰拉德（ZELDA FITZGERALD）

I love you because you have done more than any creed could have done to make me good, and more than any fate could have done to make me happy.

我爱你，因为你为我做的一切，比任何教义更让我获益，比任何命运的安排更使我幸福。

——罗伊·克罗夫特（ROY CROFT）

Nothing in this world is more powerful than love. Not money, greed, hate or passion. Words cannot describe it. Poets and writers try. They can't because it is different for each of us.

在这个世界上，最强大的不是金钱、贪婪或仇恨，也不是激情，而是爱情。在爱情面前，任何辞藻都显得苍白。诗人和作家曾试图描画爱情，却是徒劳，因为这世上没有两片相同的树叶，也没有相同的爱情。

——无名氏（AUTHOR UNKNOWN）

Wherever you're beside me... that's my home.

只要你在我身边……四海皆为家。

——比利·乔尔（BILLY JOEL）

You smile in passing, touch my shoulder. I walk with you in the garden, sharing the last of the light, the flickering of bats, the scent of roses. We are at home in quietness. Passion and the everyday flow from each other – equal expressions of our love.

你微笑着走过，轻柔地碰触我的肩膀。我与你在花园漫步，分享着夕阳的余晖，蝙蝠起舞，玫瑰芬芳。我们静静地在家，不论是激情还是平凡的生活，都代表了我们的爱情。

——夏洛特·格雷（CHARLOTTE GRAY）

Love, you know, is strangely whimsical, containing affronts, jabs, parleys, wars then peace again. Now, for you to ask advice to love by, is as if you ask advice to run mad by.

你是知道的，爱情很奇妙，爱情有冒犯，有攻击，还有谈判和战争，而最终却又归于平和。现在，你来寻求爱情的忠告，这无异于求教如何使自己发疯。

——特伦斯（TERENCE）

As selfishness and complaint pervert and cloud the mind, so love with its joys clears and sharpens the vision.

自私和抱怨使心灵扭曲、阴暗，愉悦的爱情则使视野明朗、开阔。

——海伦·凯勒（HELEN KELLER）

No one who has ever brought up a child can doubt for a moment that love is literally the life-giving fluid of human existence.

但凡抚育过孩子的人，都不会怀疑"爱"的确是人类存在的生命之源。

——斯迈利·布兰顿（SMILEY BLANTON）

That which is loved is always beautiful.
沐浴爱情的人，永远美丽。

——挪威谚语（NORWEGIAN PROVERB）

To see her is to love her, and love but her forever, for nature made her what she is, and ne'er made another!

谁见到她都会对她一见倾心、永远爱她，因为她就是天生的可人儿，无与伦比！

——罗伯特·彭斯（ROBERT BURNS）

Love is a taste of paradise.
爱情就是天堂。

——肖洛姆·阿莱汉姆（SHOLEM ALEICHEM）

Love like yours creates a world within a world. A refuge. Somewhere to call home.

爱，好比是你在一个世界中创造出来的另一个世界。爱是避难所，是一个可以称为"家"的地方。

——帕姆·布朗（PAM BROWN）

Carry me off into the blue skies of tender loves, roll me in dark clouds, trample me with your thunderstorms, break me in your angry rages. But love me, my adored lover.

你把我带到充满柔情爱意的蓝天，却把我卷进乌云，用你的雷暴伤害我，在你的盛怒中，我支离破碎。但是，我崇拜的爱人啊，爱我吧！

——萨拉·伯恩哈特（SARAH BERNHARDT）

Love is a peasant emotion and thrives as well in stables as in palaces. Of all the errands life seems to be running, of all the mysteries that enchant us, love is my favorite.

爱情是一种农夫式的情感，无论是在马厩里还是宫殿，爱情都能茁壮生长。在人生的使命中，在一切让我们着迷的神秘事物中，我最钟情于爱情。

——黛安娜·阿克曼（DIANE ACKERMAN）

Love is a force more formidable than any other. It is invisible – it cannot be seen or measured – yet is powerful enough to transform you in a moment, and offer you more joy than any material possession ever could.

爱情比任何其他力量更强大。爱情无形——看不见也无法度量——却足以强大到瞬间让你脱胎换骨，并赋予你任何物质都无法比拟的快乐。

——芭芭拉·德·安杰利斯（BARBARA DE ANGELIS）

...No one knows how it is that with one glance a boy can break through into a girl's heart.

……无人知晓，一个男孩如何仅仅用一个眼神就打动了一个女孩的芳心。

——拿破仑·波拿巴（NAPOLEON BONAPARTE）

He poured so gently and naturally into my life like batter into a bowl of batter. Honey into a jar of honey. The clearest water sinking into sand.

他就这样既温柔又自然地闯进了我的生活，就像面糊滑入碗里、蜂蜜流进蜜罐中、最清澈的水浸入沙子里。

——贾斯廷·悉尼（JUSTINE SYDNEY）

Love blossoms when there is just the right amount of tenderness combined with a long leash.

只要有足够的温柔和自由，爱情之花就会盛开。

——琼·安德森（JOAN ANDERSON）

Women wish to be loved without a why or a wherefore; not because they are pretty, or good, or well-bred, or graceful, or intelligent, but because they are themselves.

女人希望爱她们不需要任何理由。不是因为她们漂亮、善良、有教养、优雅或者聪明，而仅仅因为她们是她们。

——亨利·弗雷德里克·埃米尔（HENRI FRÉDÉRIC AMIEL）

Date: / /

Mon.	Tue.	Wed.	Thu.	Fri.	Sat.	Sun.

You are the future of my past, the present of my always, the forever of my now.

你是我过去的未来，我永远的现在，我现在的永远。

——查尔斯·吉格纳（CHARLES GHIGNA）

All that I love loses half its pleasure if you are not there to share it.

如果没有你与我分享这一切，我所爱的一切都将失去一半的欢愉。

——克拉拉·奥尔特加（CLARA ORTEGA）

283

Every day I know you better. Every day I love you more.
我对你的了解与日俱增，我对你的爱日渐浓烈。

——帕姆·布朗（PAM BROWN）

It's easy to forget that what people need most in the world cannot be satisfied by material things alone. What they ask of us is love.

我们轻易就会忘记，人们在世上最需要的东西并非仅仅用物质就能满足。人们需要的是爱。

——摘自《弗朗西斯·盖伊的友谊之书》
(FROM "THE FRIENDSHIP BOOK OF FRANCIS GAY")

Love is like young rice: transplanted, still it grows.
爱情就像水稻的幼苗，改变环境仍然可以茁壮成长。
——《马达加斯加的爱情报价单》（MADAGASCAN LOVE QUOTATION）

A life without any love is no life at all.
没有任何爱的生活根本就不是生活。

——凯瑟琳·弗莱特（KATHRYN FLETT）

When love enters into people whose hearts are not withered, it makes them want to love everyone.

当爱情走进尚未枯萎的心灵，它就会使人们想去爱每一个人。

——西蒙娜·德·波伏娃（SIMONE DE BEAUVOIR）

Where love is concerned, too much is not ever enough!

就爱情而言，再多也不够！

——皮埃尔·奥古斯汀·卡伦·德·博马舍
（ PIERRE AUGUSTIN CARON DE BEAUMARCHAIS ）

Love with all your heart. Listen with all your soul.
全心全意去爱，用整个灵魂去倾听。
——斯图尔特与琳达·麦克法兰（STUART & LINDA MACFARLANE）

One word frees us of all the weight and pain of life: The word is "love."

有一个词能让我们所有人摆脱人生的重负与痛苦：这个词就是"爱"。

——索福克勒斯（SOPHOCLES）

Date:　　/　　/

Mon.	Tue.	Wed.	Thu.	Fri.	Sat.	Sun.

Without love there is no life.

若无爱情，何谓生活。

——托马斯·马萨里克（THOMAS MASARYK）

All those millions of lives – and yet we found one another.

茫茫人海中，我们仍然找到了彼此。

——帕姆·布朗（PAM BROWN）

Love is when the desire to be desired takes you so badly that you feel you could die of it.

渴求被对方需要，这一念头热切得使你痛苦，这就是爱情。

——亨利·德·图卢兹－罗特列克（HENRI DE TOULOUSE-LAUTREC）

There is nothing more lovely in life than the union of two people whose love for one another has grown through the years from the small acorn of passion to a great rooted tree.

生命中最可爱的事，莫过于和相爱的人结合，他们的爱恋与日俱增，爱情已经从一粒小小的橡树种子成长为一棵根深叶茂的大树。

——维塔·萨克维尔－韦斯特（VITA SACKVILLE-WEST）

Thank you for believing I'm special.
谢谢你相信我很特别。

——帕姆·布朗（PAM BROWN）

The countless generations like Autumn leaves go by: Love only is eternal,
love only does not die....

无数代人犹如秋叶逝去：唯有爱情永恒，唯有爱情不朽……

——哈里·肯普（HARRY KEMP）

A heart enlightened by love is more precious than all of the diamonds and gold in the world.

一颗被爱情照亮的心，胜过世间所有的珠宝黄金。

——穆罕默德·阿里（MUHAMMAD ALI）

Love changes things; it is the most powerful force in the world. A person motivated by love is the most potent force there is. There are other forces that are potent, such as hatred. But love is a greater force.

爱情是世界上最强大的力量，爱情可以改造一切，有爱情激励的人最强大。仇恨的力量也很强大，但是，爱情的力量更伟大。

——米勒德·富勒（MILLARD FULLER）

Without love our life is a ship without a rudder… like a body without a soul.
无爱的人生犹如无舵的航船……仿佛一具没有灵魂的躯壳。

——肖洛姆·阿莱赫姆（SHOLEM ALEICHEM）

In love,
Dare to dream,
For two shape a future,
Filled with happiness,
Beyond imagination.

恋人们敢于梦想，
因为两个人共创未来，
幸福满满，
超乎想象。

——斯图尔特与琳达·麦克法兰（STUART & LINDA MACFARLANE）

To love and to be loved is the greatest happiness of existence.

爱与被爱是人生最大的幸福。

——悉尼·史密斯（SYDNEY SMITH）

Love turns one person into two and two into one.
爱情把一个人变成两个人，但两个人却有同一颗心。

——艾萨克·阿巴伯内尔（ISAAC ABARBANEL）

This is the life I love. For I'm with you. You are the root of all I do, and am.
这就是我热爱的生活。我所做的一切是因为有你与我同在，因为你，我成为我。

——帕姆·布朗（PAM BROWN）

…were I crowned the most imperial monarch,
Thereof most trustworthy, were I the fairest youth
That ever made eye swerve, had force and knowledge
More than was ever man's.
I would not prize them
Without her love.

……就算我被加冕为最高贵的君王，
由此最受人信赖；
纵然我就是那最引人注目的青年才俊，
拥有人类最强大的力量和最丰富的学知。
但是，如果没有她的爱情，
这些又怎能令我动容。

——威廉·莎士比亚（WILLIAM SHAKESPEARE）

Only when one is open to receive and absorb love can it occur.

只有当你敞开心扉去接受和吸收爱情，爱情才会发生。

——琼·安德森（JOAN ANDERSON）

Until the stars fall from the sky,
Until the seas melt away,
I shall always love you.

星辰陨落，
沧海枯竭，
我爱你的心永不变。

——琳达·麦克法兰（LINDA MACFARLANE）

Romance is fine, and passion too. But best of all is knowing you are there.

浪漫和激情都令人陶醉。但是，最重要的是，我知道你就在那里。

——珍妮·德·弗里斯（JENNY DE VRIES）

Love is a game that two can play and both win.

爱情是一场只需两个人就能玩的双赢游戏。

——伊娃·伽柏（EVA GABOR）

Life is the first gift, love is the second, and understanding the third.

生命是第一份礼物，爱情是第二份，第三份则是理解。

——玛吉·皮尔希（MARGE PIERCY）

Love is knowing someone else cares.

爱情就是知道有人在乎你。

——无名氏（AUTHOR UNKNOWN）

One hour of right-down love is worth an age of dully living on.
与其平淡一生，不如投入地爱一次，哪怕只是短暂的一小时。

——阿芙拉·贝恩（APHRA BEHN）

Union gives strength.
团结会产生力量。

——《伊索寓言》（AESOP'S FABLES）

Love distracts one from the tidiest plans, the narrowest course, the clearest goals.
爱情使人变得没有条理，爱情扰乱了我们最精确的航向，最明确的目标。

——黛安娜·阿克曼（DIANE ACKERMAN）

A very ordinary couple in a very ordinary house – but what an extraordinary love we share.

在一所非常普通的房子里，住着一对非常平凡的爱侣——但是，我们分享着多么不平凡的爱情。

——克拉拉·奥尔特加（CLARA ORTEGA）

There are no limits to love.

爱情无极限。

——迈克尔·凯恩（MICHAEL CAINE）

To be loved is the triumph of living.
被爱就是人生的胜利。

——安东尼·奎因（ANTHONY QUINN）

Love...can only be given and received. It cannot be taken.

爱情……只能是给予和接受，你无法偷走它。

——奥莉亚·山居梦客（ORIAH MOUNTAIN DREAMER）

Being in love is the excitement of the moment, that sharp blade of exhilaration which cleaves through the seconds and makes the world sparkle.

坠入爱河是瞬间的兴奋，是穿透时间的利刃，爱情让世界熠熠生辉。

——史蒂夫·鲍凯特（STEVE BOWKETT）

Nothing is more important to human beings than to be loved.
对人类而言，最重要的事情莫过于被爱。

——西莉亚·鲍林（CELIA BOWRING）

A good marriage is passion and monotony, practicalities, magic, talk and tears and laughter. And at the core lies a secret place. A place of trust and love and deep content.

一段美好的婚姻既有激情也有平淡，既实用，也神奇，有谈笑风生也有泪水。在婚姻的深处有一个神秘的地方，这是一个充满信任、爱和美满的地方。

——帕姆·布朗（PAM BROWN）

Love is comfort in sadness, quietness in tumult, rest in weariness, hope in despair.

爱情是悲伤时的慰藉、喧嚣中的宁静、疲惫时的休憩、绝望时的希望。

——玛丽昂·C. 盖瑞提（MARION C. GARRETTY）

You make me glad to be alive.
你使我乐于活着。

——斯图尔特与琳达·麦克法兰（STUART & LINDA MACFARLANE）

Man and woman are two locked caskets, of which each contains the key.
男人和女人是两个上了锁的匣子，里面锁着那把开锁的钥匙。

——伊萨克·丹森（ISAK DINESEN）

Love is like playing the piano. First you must learn to play by the rules, then you must forget the rules and play from your heart.

爱情，犹如弹钢琴。首先，你必须学会遵守弹奏规则，然后，你又必须忘记这些规则，用心去演奏。

——无名氏（AUTHOR UNKNOWN）

Love is trusting, accepting and believing, without guarantee. Love is patient and waits, but it's an active waiting, not a passive one. For it is continually offering itself in a mutual revealing, a mutual sharing. Love is spontaneous and craves expression through joy, through beauty, through truth, even through tears. Love lives the moment; it's neither lost in yesterday nor does it crave for tomorrow. Love is Now!

爱是毫无保留的信任、接受和信仰；爱需要耐心等待，而不是被动守候；爱是相互倾诉、相互分享。爱自然而然，于欢乐、美丽、真理，甚至眼泪中我们都能看到爱。爱就在当下，爱既不迷失在昨天，也不渴求明天。爱就是现在！

——利奥·布斯卡利亚（LEO BUSCAGLIA）

We have made a home, gathered possessions, kept treasured souvenirs of all the joys we've shared. All dear to me – yet if all were lost tomorrow and we were safe in each other's love – I'd mourn them only for a moment. All that is most precious can be held in arms.

我们建造了一个家，收集了财物，珍藏了那些记录着我们曾共同享有的所有欢乐时光的纪念品。所有这一切对我都很珍贵——然而，如果明天这一切都逝去，而我们却在彼此的爱情中安然无恙——那么，我只会为这些失去的东西悲伤片刻。最重要的是，我们仍然可以把最珍贵的东西拥入怀中。

——帕姆·布朗（PAM BROWN）

Love is, above all, the gift of oneself.
归根结底，爱是一个人的天赋。

——让·阿努伊（JEAN ANOUILH）

I have spread my dreams under your feet; Tread softly because you tread on my dreams.

我已经把梦想铺展在你的脚下，请温柔一点吧，因为你的脚踩着我的梦想。

——威廉姆·巴特勒·叶芝（WILLIAM BUTLER YEATS）

We need to be loved and to love.

我们需要爱与被爱。

——奇夫·丹·乔治（CHIEF DAN GEORGE）

We treasure love. It quenches, vexes, guides, and murders us. It seeps into the mortar of all our days. It feeds our passion, it fills our fantasies, it inspires our art.

我们珍视爱情。它时而熄灭我们的希望，时而令我们烦恼，时而引导我们，时而打垮我们，爱情可以渗透每一个如灰泥般沉重的日子。爱情滋养了我们的激情，充盈着我们的幻想，激发灵感。

——黛安·阿克曼（DIANE ACKERMAN）

Whatever our souls are made of, his and mine are the same.
不管我们的灵魂是用什么做成的，我和他拥有相同的灵魂。
——埃米莉·勃朗特（EMILY BRONTË）

And true love holds with gentle hands the hearts that it entwines.
真爱是用温柔的双手呵护那两颗彼此缠绕的心。

——无名氏（AUTHOR UNKNOWN）

Congratulations! You found each other! Out of millions of lives – scattered across the planet – against all odds, all possibilities, all chance you found each other. Never let each other go.

祝贺你们，你们找到了彼此！尽管困难重重，充斥着各种变数、各种可能，但是你们仍然穿越茫茫人海，从大千世界中找到了彼此，从此永不分离。

——帕姆·布朗（PAM BROWN）

Love is born with the pleasure of looking at each other, it is fed with the necessity of seeing each other, it is concluded with the impossibility of separation!

爱情天然就包含着彼此欣赏的乐趣，长相守才能滋养爱情，永远不离弃！

——何塞·马蒂·佩雷斯（JOSÉ MARTÍ Y PERÉZ）

Your breath, like a spring shower makes my body tingle. Your touch, like the summer sun makes my temperature soar. Your voice, like an autumn breeze sets my heart a flutter. Your kiss, like the winter snow sends shivers through my soul.

你的呼吸，犹如绵绵春雨，滋润着我的身体；你的爱抚，犹如夏日骄阳，沸腾着我的热血；你的声音，犹如秋日微风，荡漾着我的心扉；你的香吻，犹如冬日白雪，颤动着我的灵魂。

——琳达·麦克法兰（LINDA MACFARLANE）

You can't go out and earn love, you can't buy it, borrow it or even look for it. You can look for a date and sex, but not true love. It doesn't work like that.

爱情赚不来，买不到，借不着，也寻不得。你可以去约会，可以寻欢作乐，但无法找到真爱，这些招数在真爱面前可不奏效。

——黛西·杜利（DAISY DOOLEY）

If you ever leave me, I'm coming with you.

如果有一天你要离我而去，我就和你一起走。

——娜奥米·贾德（NAOMI JUDD）

Love is the common denominator that goes through all cultures and binds us together. Without it, we're lost.

爱情的共性是超越文化的障碍。如果没有爱情，我们就会迷失自己。

——熊心（BEARHEART）

Date: / /

Mon.	Tue.	Wed.	Thu.	Fri.	Sat.	Sun.

I spread out the days before me – the days we have spent together – some bright as stars, some glowing with an opalescent magic, some cool as pebbles, some flashing with fire. What would they have been without you? Dates crossed off on a calendar.

回首我们曾一起走过的岁月，有些灿若朗星，有些泛着珠光幻彩，有些如鹅卵石般清凉，有些则闪烁着火焰的光芒。可是，如果没有你，这些岁月会是怎样一副模样？它们不过是日历上逐个画掉的数字罢了。

——玛丽昂·C.加勒提（MARION C. GARRETTY）

一般而言，亲密的感情是一首灵魂之歌，共邀两人分享彼此的灵魂，它是一首无人能够抗拒的歌。

Intimacy in general terms is a song of spirit inviting two people to come and share their spirit together. It is a song that no one can resist.

一般而言，亲密的感情是一首灵魂之歌，共邀两人分享彼此的灵魂，它是一首无人能够抗拒的歌。

——索本夫·索蒙（SOBONFU SOMÉ）

Love from one being to another can only be that two solitudes come nearer, recognize and protect and comfort each other.

只有当两颗孤独的心彼此靠近、认可，彼此呵护、安慰，才会产生爱情。

——韩素音（HAN SUYIN）

Communication from a heart to a heart, a soul to a soul, is worth more than a whole page, a whole volume of books that could be written.

心与心的交流，灵魂与灵魂的沟通，比写一整页纸、一整本书更有价值。

——马斯科吉部落（THE MUSKOGEE TRIBE）

The love of ordinary people lights the world.
平凡人的爱情足以照亮整个世界。

——珍妮·德·弗里斯（JENNY DE VRIES）

Love doesn't mean never having to say you are sorry. Love, at least lasting love, means being able to recognize when you need to say you are sorry.

爱情并不意味着永不抱歉。爱情，那持久的爱情，意味着你知道自己何时需要抱歉。

——卡门·勒妮·贝里与塔玛拉·特瑞德
(CARMEN RENEE BERRY & TAMARA TRAEDER)

Love is not getting, but giving. Love is the best thing in the world and the thing that lives the longest.

爱情不是索取，而是给予。这世上爱情最美好、最长久。

——亨利·凡·戴克（HENRY VAN DYKE）

How sad and bad and mad it was – But then, how it was sweet!

爱情多么令人悲伤，多么使人沮丧，多么让人发狂——然而，它又是多么甜蜜啊！

——罗伯特·布朗宁（ROBERT BROWNING）

Love cannot be regulated or sustained by structures, rules or commitments. It can only be sustained by continuing acts of love which are marked by gentleness, care, openness and trust.

我们无法依靠组织、规则和承诺来安排和维系爱情，爱情是持久的温柔、关爱、坦诚和信任。

——菲奥娜·卡斯尔（FIONA CASTLE）

Love is the transforming power in our human nature. Love cries for life. Love fights for life.

爱是人性中的变革力量。爱呼唤生命，也为生命而战斗。

——钦莫伊（SRI CHINMOY）

Winter twilight
On a window pane
I write your name.

冬日的黄昏，
在窗玻璃上，
我写下你的名字。

——日本俳句（JAPANESE WISDOM）

Nothing can transform one's life so dramatically as love breaking through. It is like the sun bursting through dark clouds. Suddenly everything is amazingly different.

没有什么比邂逅爱情更能明显地改变一个人的人生了，仿佛太阳冲破乌云，刹那间，一切迥然不同。

——温迪·克雷格（WENDY CRAIG）

Falling in love is the greatest excitement of all.

坠入爱河最激动人心。

——帕姆·布朗（PAM BROWN）

To love is to forgive, to understand, to wish the other's happiness....

爱他就是去原谅他、理解他，并且希望他幸福……

——阿娜伊斯·宁（ANAÏS NIN）

They are closest to us who best understand what life means to us, who feel for us as we feel for ourselves, who are bound to us in triumph and disaster, who break the spell of our loneliness.

爱情最靠近我们，爱情最了解生命对我们意味着什么，感知爱情就如同感知自己；不论是胜利还是灾难，爱情都与我们紧紧相依，爱情打破了我们孤独的魔咒。

——亨利·阿朗佐·迈尔斯（HENRY ALONZO MYERS）

We live by admiration, hope and love…especially love.

我们依靠赞赏、希望以及爱情而生活……尤其是依靠爱情而生活。

——摘自《弗朗西斯·盖伊的友谊书》
(FROM "THE FRIENDSHIP BOOK OF FRANCIS GAY")

Love is something that you can leave behind when you die. It's that powerful.

当你撒手人寰，而爱仍然永存。爱就是这么强大的力量。

——约利克（JOLIK）

I never see beauty without thinking of you or scent happiness without thinking of you. You have fulfilled all my ambition, realized all my hopes, made all my dreams come true.

一想到你，我就会看到美丽，闻到幸福的气息。你让我实现所有的抱负和希望，你让我一切美梦成真。

——艾尔弗雷德·达夫·科珀爵士（SIR ALFRED DUFF COOPER）

357

Immature love says: "I love you because I need you."
Mature love says: "I need you because I love you."

幼稚的爱情会说："我爱你，因为我需要你。"而成熟的爱情则会说："我需要你，因为我爱你。"

——埃里克·弗罗姆（ERICH FROMM）

Who would not welcome anyone whose heart is filled with nothing but love?

谁会不喜欢心中只有爱的人呢?

——维诺巴·比哈夫（VINOBA BHAVE）

Never lose a chance to tell someone you love them!
不要放弃任何机会去告诉他，你爱他！

——摘自《弗朗西斯·盖伊的友谊书》
（ FROM "THE FRIENDSHIP BOOK OF FRANCIS GAY" ）

All my soul follows you, love – encircles you – and I live in being yours.

我全心全意地追随你、爱你、环绕着你，我活在你的世界里。

——罗伯特·布朗宁（ROBERT BROWNING）

I love thee to the depth and breadth and height
My soul can reach….
I love thee with the breath,
Smiles, tears, of all my life!

我爱你，直至灵魂所能触及的深度、宽度和高度……
我用毕生的呼吸、微笑与泪水爱你！

——伊丽莎白·芭蕾特·布朗宁（ELIZABETH BARRETT BROWNING）

Love changes everything.
爱情能够改变一切。

——琳达·麦克法兰（LINDA MACFARLANE）

I can face anything with my hand in yours.
握紧你的手，我就能面对一切。

——帕姆·布朗（PAM BROWN）

The only thing we'll be remembered for when we die is the love we leave behind.

辞别人世后，能被人们记起的唯有我们留下的爱。

——伊丽莎白·库伯勒·罗斯（ELISABETH KÜBLER-ROSS）

'Tis love, 'tis love, that makes the world go round!

是爱，是爱让世界转动！

——刘易斯·卡罗尔（LEWIS CARROLL）

365 days of love: The greatest love quotes of all time
Published in 2018 by Helen Exley Giftbooks in Great Britain.
illustrated by Julitte Clarke © Helen Exley Creative Ltd 2011.
Selection and arrangement by Dalton Exley © Helen Exley Creative Ltd 2011.
Dedicated, with all my love, to Lisa Fajembola.

The Simplified Chinese translation rights are arranged through RR Donnelley Asia
Simplified Chinese edition copyright: 2020 New Star Press Co., Ltd.
All rights reserved.
著作版权合同登记号：01-2019-4971

图书在版编目（CIP）数据

我怎么能够把你来比作夏天：爱情手账：汉英对照 /（英）多尔顿·埃克斯利 (Dalton Exley) 编选；周成刚主编 . -- 北京：新星出版社，2020.11
（新新悦读）
ISBN 978-7-5133-3733-5

Ⅰ . ①我… Ⅱ . ①多… ②周… Ⅲ . ①本册 ②格言－汇编－世界－汉、英
Ⅳ . ① TS951.5 ② H033

中国版本图书馆 CIP 数据核字 (2020) 第 191615 号

我怎么能够把你来比作夏天：爱情手账

编　　选：[英] 多尔顿·埃克斯利 (Dalton Exley)
丛书主编：周成刚

策划编辑：李金学　赵　丹
责任编辑：汪　欣
特约编辑：张翔宇　李潇潇
责任校对：刘　义
责任印制：李珊珊
装帧设计：冷暖儿

出版发行：新星出版社
出 版 人：马汝军
社　　址：北京市西城区车公庄大街丙3号楼　　　100044
网　　址：www.newstarpress.com
电　　话：010-88310888
传　　真：010-65270449
法律顾问：北京市岳成律师事务所

读者服务：010-88310811　　service@newstarpress.com
邮购地址：北京市西城区车公庄大街丙3号楼　　　100044

印　　刷：北京尚唐印刷包装有限公司
开　　本：889mm×1194mm　　1/32
印　　张：11.75
字　　数：50千字
版　　次：2020年11月第一版　　　2020年11月第一次印刷
书　　号：ISBN 978-7-5133-3733-5
定　　价：78.00元